Technology Guide

to accompany

Chemistry, Sixth Edition

Steven S. Zumdahl
Susan A. Zumdahl

Contents

Logging on to the *Chemistry, Sixth Edition* Web Site	1
Logging on to SMARTHINKING™ Online Tutoring	2
Launching and Using **General Chemistry Interactive 6.0** Student CD-ROM	3
Chime Introduction	6
Media Activities	9
License Agreement	24

Vice President and Publisher: Charles Hartford
Executive Sponsoring Editor: Richard Stratton
Instructional Technology Sponsor: Susan Warne
Development Editors: Sara Wise, Rita Lombard
Editorial Assistant: Rosemary Mack
Senior Marketing Manager: Katherine Greig
Marketing Associate: Alexandra Shaw
Internet Producer: Jennifer Yuan
Software Producer: Chere Bemelmans

This CD-ROM contains Macromedia Shockwave™ Player and Macromedia Flash™ Player software by Macromedia, Inc., Copyright © 1995-2000 Macromedia, Inc. All rights reserved. Macromedia, Shockwave, and Flash are trademarks of Macromedia, Inc.

QuickTime and the QuickTime logo are trademarks used under license. The QuickTime logo is registered in the U.S. and other countries.

Copyright © 2003 by Houghton Mifflin Company. All rights reserved.

No part of this work may be reproduced or transmitted in any form or by any means, electronic or mechanical, including photocopying and recording, or by any information storage or retrieval system without the prior written permission of Houghton Mifflin Company unless such copying is expressly permitted by federal copyright law. Address inquiries to College Permissions, Houghton Mifflin Company, 222 Berkeley Street, Boston, MA 02116-3764.

Printed in the U.S.A.

Logging on to the *Chemistry, Sixth Edition* Web Site

1. Go to chemistry.college.hmco.com/students.
2. Choose the *Zumdahl Chemistry 6/e* website link by clicking on the cover of the book or choosing the title in the drop-down "General" list.
3. The user name is "acid" and the password is "base."

The web site for *Chemistry* offers a robust range of resources to enhance your study of general chemistry, including:

- Houghton Mifflin's ACE Practice Tests: Three quizzes per chapter. ACE provides immediate feedback on your results, plus references to the sections in the textbook for further study.
- Additional Tutorials and Movies to supplement those found on the CD-ROM enclosed in this package
- Flashcards of key terms and concepts
- An interactive periodic table, with self-quiz questions
- *Chemical Impact* boxes
- A library of molecules viewable through Chime software
- Links to chemistry resources on the Web
- Links to **SMARTHINKING**™ online tutoring.

Logging on to **SMARTHINKING**™ Online Tutoring
from the student web site

1. Click on **SMARTHINKING**™
2. Peel off the sticker below, to get your username and password.
3. Follow the instructions online to set up your student account.

> **To set up your SMARTHINKING online tutoring account:**
>
> Go to
> http://www.smarthinking.com/houghton.html
>
> Use this username and password to create your account.
>
> **Username:** HM19438281507
>
> **Password:** HM1973

If you prefer, it is also possible to access SMARTHINKING ™ tutoring directly from smarthinking.com/houghton.html

For technical support for SMARTHINKING™*, call (888) 688-7560, extension 4 or e-mail* support@smarthinking.com.

SMARTHINKING™ is our free, live online tutoring service. It allows access to support from a real professor or study resources from wherever you are, whenever you need help. With SMARTHINKING™ you can:

- Connect immediately to live help Sunday-Thursday, 2pm-5pm and 9pm-1am EST
- Submit a question to get a response from an e-structor, usually within 24 hours
- Use the whiteboard with full scientific notation and graphics
- Pre-schedule time with an e-structor
- View past online sessions, questions, or essays in an archive on your personal academic homepage
- View your tutoring schedule
- Work on other projects while waiting for help

SMARTHINKING™ offers features and benefits to help you attain the best possible grade.

Launching and Using General **Chemistry Interactive 6.0** Student CD-ROM

*The following pages explain how to launch and run **General Chemistry Interactive 6.0**. We assume that you are already familiar with the conventions of Macintosh operating systems or Microsoft Windows. If you need more information, please consult those respective manuals. After launching the program, please consult Help in the menu if you need more information on navigation or functionality.*

Minimum Windows® System Requirements
- Microsoft® Windows® 95, 98, NT 4.0, ME, 2000, XP
- Pentium Processor or compatible
- 32MB RAM
- Display adapter set for thousands of colors or better
- 800 x 600 resolution
- Windows-compatible sound card
- Speakers or headphones
- Four-speed CD-ROM drive
- A multimedia PC level II is recommended
- Netscape Navigator 4.0 and higher OR
- Microsoft Internet Explorer 4.0 or higher
- QuickTime® 5.0 (included)
- Shockwave Player (included)

General Chemistry Interactive 6.0 for Windows®
1. Insert **General Chemistry Interactive 6.0** into your disk drive and close the door.
2. From Windows Explorer or My Computer, double-click the file start.html on the root level of your **General Chemistry Interactive 6.0** CD-ROM.

From the main menu, select a chapter to begin. For more information on how to use **General Chemistry Interactive 6.0**, see the file How to Use the CD on the root directory of the CD, or click Help, available from within **General Chemistry Interactive 6.0**.

Note: QuickTime for Windows 5.0 and Shockwave 8 must be installed on your computer.

To install QuickTime for Windows 5.0 and the Shockwave Player:
1. Locate the QuickTime installer—QuickTime.exe—on **General Chemistry Interactive 6.0** in the folder QuickTime. Double-click on QuickTime.exe and follow the on-screen instructions.

2. Locate the Shockwave installer—Shockwave_851_Installer.exe—in the Shockwave folder on **General Chemistry Interactive 6.0**. Double-click the Shockwave installer and follow the on-screen instructions.

Note: If you cannot see the videos or animations in the Visualizations, and you have installed QuickTime and Shockwave from this disk, you may have a problem with the number of versions of Shockwave installed on your computer.

1. From **General Chemistry Interactive 6.0**, go to the QT3Asset.x32 directory and copy the file QTXtra.32.
2. On your hard drive, using Windows Explorer or My Computer, go to Windows\System\Macromed\Shockwave\Xtras and paste the file. If there is more than one Shockwave directory within the Macromed directory (Shockwave, Shockwave 8, etc,) paste the file in each Shockwave\Xtras directory. (Note: on Windows NT, the path to the Xtras folder is Winnt\System32\ Macromed\Shockwave\Xtras.)

Minimum Macintosh® System Requirements
- MAC OS 8.6 or higher
- Power PC processor or compatible
- 32 MB RAM
- Display adapter set for thousands of colors or better
- 800X600 resolution
- Four-speed CD-ROM
- Speakers or headphones
- Netscape Navigator 4.0 and higher OR
- Microsoft Internet Explorer 4.0 or higher
- Shockwave Player (included)
- QuickTime™ 5.0 (included)

Chemistry Interactive 6.0 for Macintosh®
1. Insert **General Chemistry Interactive 6.0** into your disk drive and close the door.
2. Double-click the CD-ROM icon labeled **Chemistry Interactive** on your desktop.
3. Then double-click the file start.html (IE users) or startN.html (Netscape users).

From the main menu, select a chapter to begin. For more information on how to use **General Chemistry Interactive 6.0**, see the file How to Use the CD on the root directory of the CD, or click Help, available from within **General Chemistry Interactive 6.0**.

Note: QuickTime 5.0 and Shockwave 8 must be installed on your computer.

To install QuickTime 5.0 and the Shockwave Player:

1. Locate the QuickTime installer (filename QuickTime Installer) in the folder QuickTime Installer on **General Chemistry Interactive 6.0**. Double-click QuickTime Installer and follow the on-screen instructions.
2. Locate the Shockwave installer (filename Shockwave Installer) in the folder Shockwave Installer on **General Chemistry Interactive 6.0**. Double-click Shockwave Installer and follow the on-screen instructions.

Note: If you cannot see the videos or animations in the Visualizations, and you have installed QuickTime and Shockwave from this disk, you may have a problem with the number of versions of Shockwave installed on your computer.

1. From **General Chemistry Interactive 6.0**, go to QTXtra folder and copy the file QuickTime Asset.
2. On your hard drive, go to System Folder:Extensions:Macromedia:Shockwave:Xtras and paste the file. If there is more than one Shockwave folder within the Macromedia folder (Shockwave, Shockwave 8, etc.), paste the file in each Shockwave:Xtras folder.

If you experience any problems launching or running General Chemistry Interactive 6.0, refer to the README file for hints on improving the performance of your computer.

Program Help

Help is available by clicking on the Help button at the bottom of the screen. See also the file How to Use this CD on the root directory of the CD.

Technical Support: 1-800-732-3223

For technical assistance, call Houghton Mifflin Software Support, Monday through Friday, between 9 a.m. and 5 p.m. E.S.T. Software Support can also be reached by email at support@hmco.com.

If the product is defective, contact Houghton Mifflin within 30 days of purchase at the number above. You will receive shipping instructions for the return and replacement of the defective disc(s). Provided the disc(s) has not been physically damaged, Houghton Mifflin will replace the disc(s).

Chime Introduction

Molecular structures are three-dimensional objects that have width, height and depth. It is difficult to understand spatial relationships between the various parts of the molecule using a flat two-dimensional image on a printed page, so chemists often use computers to portray these relationships. Although the image on the screen is still two dimensional, turning the molecule lets you form a three-dimensional visual impression of the molecule. One of the most commonly used internet plug-ins for creating and viewing molecular structures within a web page is Chemscape Chime.

Chime works within an internet browser to display molecular structures on your computer screen. Using your mouse, you can rotate the molecule and view it from practically any angle. You can also change the color and appearance of the entire molecule, or segments of the molecule, so that you can view more interesting aspects of the molecular structure. Advanced features of Chime were designed to display proteins and other biological polymers. Nevertheless, the program is versatile enough to make it extremely useful for displaying small inorganic and organic molecules.

Obtaining Chime: You can download Chime from the MDL Information Systems web site at: **http://www.mdlchime.com/chime** (the same information can be accesssed via the home page **http://www.mdli.com/cgi/dynamic/welcome.html**.) Instructions on how to install this plug-in can be obtained from the software company that designed your internet browser. Note that you may have to upgrade your current browser to the most recent version in order for Chime to run properly.

Finding Chime Sites*:* Although it is certainly possible to create your own web pages to display molecular structure using Chime, it is best to start with existing web pages. Many Chime pages can be found on the web site for your textbook, including a Chime tutorial with introductory examples. Once you load a web page containing a Chime window into your browser, you should see a molecular structure on your screen. This structure is contained within a separate window, referred to as the Chime window. The mouse cursor, or pointer, becomes active within this window. Start with a small molecule or a small protein when first learning how to use Chime.

Using Chime: *Default Options:* Upon opening a Chime window, you will note that the molecule may be rotated quite easily by holding down the mouse button and dragging the mouse (Mac) or holding down the left mouse button and dragging the mouse (PC). Many more useful features of Chime become accessible through use of a pop-up menu activated by holding down the mouse button while the cursor is in the Chime window (Mac) or clicking the right mouse button while the cursor is in the Chime window (PC). Once the pop-up menu is open the user has access to a wide number of customization options. Only some of these are discussed below. As with any software, if you have a question about what a particular option does, the best approach is to select it and find out.

Altering the position, orientation, or size of the molecule: You can move (translate), rotate, or resize the molecule using various mouse and keyboard combinations. For a convenient list of mouse click and keyboard combinations that enable these actions select Mouse ⊃ Open Mouse Control Help Page from the pop-menu obtained by clicking (Mac) or right-clicking (PC) in the Chime Window. A small table similar to the one at the right will be displayed indicating options for moving, rotating, and resizing the molecule.

Action	Windows	Macintosh
MENU	Right	Hold Down
Rotate X, Y	Left	Unmodified
Translate X,Y	Ctrl-Right	Command
Rotate Z	Shift-Right	Shift-Command
Zoom	Shift-Left	Shift
Slab Plane	Ctrl-Left	Ctrl

Changing the appearance of the molecule: One of the most useful features of Chime is the ability to change the appearance of a molecule on the screen. To do this, open the pop-up menu by clicking (Mac) or right-clicking (PC) in the Chime Window. Select the Display option and you will have a choice of a Wireframe, Sticks, Ball & Stick, etc. molecule displays. If you are uncertain as to the nature of any of these, simply make the selection and watch the results.

Changing the target of operations: In addition to changing the appearance or color of the entire molecule, you can also change part of the molecule by first selecting that part and then changing its color or the way in which it is displayed. You must indicate which part of the molecule you want to operate on. To do this open the pop-up menu by clicking (Mac) or right-clicking (PC) in the Chime Window. Choose the Select option from the pop-up menu. You will find a list of selectable items near the bottom of the new pop-up menu. For example, you can select individual atoms, amino acid residues, or types of amino acids. Once a particular subset of atoms is selected, all subsequent commands are only applied to those atoms.

Making measurements and Labeling Atoms: Labels can be attached to atoms and measurements can be made of atom distances, bond angles, and torsions using the Select ⊃ Mouse Click Action and choosing one of the available options from the pop-up menu. The default measurement is bond angle. To display the bond angle in the text bar at the base of the browser window, simply click on an outside atom, the middle atom and last outside atom of the bond angle of interest. To change to, for example, bond distance measurement, choose the Distance option from the Mouse Click Action menu. Click on the atoms of interest and the atom distance will be displayed in the text bar at the base of the browser window. Note the atoms do not have to be bonded to determine the distance between them.

Other Interesting Features of Mouse Click Action: In addition to the labeling and measurement functions of Select ∋ Mouse Click Action option of the pop-up menu, some other functions of this option may be of interest. Chime allows you to change *how* the mouse buttons work by using the Mouse Click Action sub-menu, which is accessed from the Select option on the main menu.

Some of the more useful tools are listed below:

- **Toggle Atom Label** lets you place a label associated with the identified atom on the screen. Once you have selected this option, you can click on an atom to place a label for that atom on the your screen.
- **Pick Center of Rotation** is useful when you are looking at a sub-section of the molecule and you want to prevent the part you are looking at from rotating out of view.

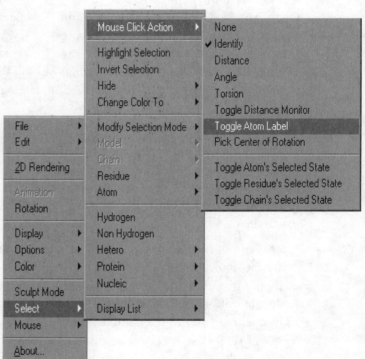

Saving your work. To save your work, you can save the molecule to the Clipboard (Edit + Copy). From there you can paste the image into reports or other homework assignments. You may want to change the background to white if you plan to paste your picture into a report or print it.

Media Activities

The Media Activities are designed to direct you to electronic resources that will help you master the material or serve as a reference. These questions will help you develop a more thorough understanding of important principles and concepts.

Chapter 1

1. Open Chapter 1 on the student CD. The initial screen for each chapter is a brief **overview** or summary of the material presented in that chapter of the text. Read the overview to focus your attention on the important topics presented in each chapter.

2. Test your understanding of Chapter 1 key terms by opening **Key Words** on the CD. For each term, come up with your own definition, and then check your definition by clicking on the specific term. Define the terms out loud with a study partner.

3. Test your understanding of Chapter 1 by taking the **ACE quizzes** on the student web site.

Chapter 2

1. Many early scientists performed experiments to determine the structure of the atom. J.J. Thomson's work with cathode-ray tubes led to the discovery of electrons. Open the Cathode Ray Tube **Visualization** to review cathode rays and the effect of a magnet on these rays.

2. There are many ways to represent structures of compounds. The Molecule Library on the student web site shows various representations of molecules referenced in each chapter. Choose CH_4 from the Molecule Library. A wireframe representation of CH_4 comes up initially; however, you can change this to other types of representations. Go to Display to see the other options. Click on each type of representation (e.g., Sticks, Spacefill, Ball and Stick) to see what they look like. You can also enhance the display by clicking Options. For each representation, try various options. Note that nothing happens when you select Labels. This option only labels atoms other then carbon and hydrogen. Since CH_4 is only composed of C and H, no labels appear. Choose some more molecules and play.

3. Test your understanding of Chapter 2 material by taking the **ACE quizzes** on the student web site.

Chapter 3

1. Open the **Visualizations** for Chapter 3 on the student CD to see two examples of very energetic chemical reactions.

(a) In the oxygen and hydrogen video, H_2 by itself is not nearly as explosive as when a mixture of H_2 and O_2 is present. Why is this? (Hint: Assume the sample containing only H_2 is reacting with O_2 in the surrounding air.)
(b) For the ammonium dichromate reaction, determine the balanced equation. Assume that ammonium dichromate is the only reactant and assume that chromium (III) oxide, nitrogen gas and water vapor are the products. Check your answer by referencing Sample Exercise 3.14 in the text.

2. Open the Conservation of Mass and Balancing Equations **Understanding Concepts** activity on the student CD to review important characteristics of chemical equations.

(a) Try the Exercises to test your ability to balance chemical equations. Click on the + or - buttons to determine the correct coefficients. You must choose the coefficients for the reactants before determining the product coefficients. The best balanced equation has smallest whole number integers. However, any chemical equation can be balanced in an infinite number of ways. Explain.
(b) Does the mole ratio between two species in a chemical equation change if smallest whole number integers are not used to balance the equation?
(c) Using the Combustion of Methane Exercise, determine some combinations of coefficients other than smallest whole number integers to balance the equation. How are the coefficients related to the smallest whole number coefficients? Pick two substances and determine the mole ratio between these substances for each possible balanced reaction. How do the mole ratios compare?

3. Open the Limiting Reactants **Understanding Concepts** activity on the student CD to examine limiting reactant problems on the microscopic level. The four exercises let you choose various numbers of reactant molecules, and then determine which is the limiting reactant. Consider the following combinations of reactant molecules that refer to the reactions in the Exercises.

(a) If 2 molecules of acetylene are reacted with 4 molecules of oxygen, what is the limiting reactant? How many H_2O molecules can form? How many molecules of O_2 are required to react completely with 2 molecules of acetylene?
(b) If 4 molecules of ammonia are reacted with 7 molecules of oxygen, what is the limiting reactant? How many NO molecules can form? How many O_2 molecules are required to react completely with 4 ammonia molecules?

(c) If 2 molecules of hydrazine are reacted with 5 molecules of oxygen, what is the limiting reactant? How many H_2O molecules can form? How many O_2 molecules are required to react completely with 2 molecules of hydrazine?
(d) If 3 molecules of methane are reacted with 5 molecules of oxygen, what is the limiting reactant? How many CO_2 molecules can form? How many molecules of O_2 are required to react completely with 3 molecules of methane? Check your answers by answering the limiting reactant question and clicking the two Hint buttons.

4. Test your understanding of Chapter 3 material by taking the **ACE quizzes** on the student web site.

Chapter 4
1. Play the Dissolution of a Solid in a Liquid **Visualization** on the student CD, which shows the hydration process for NaCl.

(a) Which color sphere represents the Na^+ ions?
(b) Which color represents the Cl^- ions?
(c) Is NaCl(aq) a strong, weak, or nonelectrolyte? Explain the differences between the various types of electrolytes.

2. Open the Conductivities of Aqueous Solutions **Visualization** and play the demonstration illustrating the differences in conductivity between strong, weak, and nonelectrolytes in water. In the demo, nothing happened to the light when distilled water was used, the bulb was very dimly lit for sparkling mineral water, the bulb was slightly brighter for the soft drinks tested, and was brightest when HCl(aq) was tested.

(a) Explain these results.

(b) What would you expect to happen if sugar water [$C_{12}H_{22}O_{11}$(aq)] was tested?

(c) Finally, summarize which compounds are generally classified as strong electrolytes? As weak electrolytes? As nonelectrolytes? Make sure you understand the illustrations in Figures 4.5-4.9 in the text, which show the differences between the types of electrolytes.

3. Open the Precipitation Reactions **Understanding Concepts** activity on the student CD and go through the introduction and example. The animation of the silver chloride is important to understand.

(a) What is the blue ball in the animation?
(b) What ion is missing from the animation?
(c) Consider the reaction between 4 formula units of $Pb(NO_3)_2$(aq) and 3 formula units of $CaCl_2$(aq). After precipitate formation is complete, how many and which type of ions remain in solution? What ion is the limiting reactant in this reaction? Understanding atomic level views of precipitation reactions is extremely important. Drawing a picture is often helpful.

4. In the simplest form, acid-base reactions involve the transfer of a proton (H^+) from an acid to a base. When the base contains OH^- ions, the H^+ from the acid reacts with OH^- to form H_2O (we commonly say that the added OH^- neutralizes the acid). In real life, acid-base reactions are more complicated than this and are discussed in more detail in Chapters 14 and 15 of the text. To see an advanced look at acid-base reactions, open the Acid/Base **Understanding Concepts** activity on the student CD. Note that this activity goes into more detail than the ideas presented on acid-base reactions in Chapter 4 of the text.

5. Oxidation-reduction reactions are another common type of aqueous reactions. Open the Oxidation-Reduction Reactions **Understanding Concepts** activity on the student CD and go through the introduction to review pertinent ideas. In the animation, what color sphere represents tin? Zinc? What are the green balls? Do the Example and Exercises only if your class has discussed the activity series.

6. Open the Zinc and Iodine **Visualization** and play the demo to see an energetic oxidation-reduction reaction. The product of the reaction is zinc iodide. With this in mind, what is oxidized? What is reduced? What is the oxidizing agent? What is the reducing agent?

7. Test your understanding of Chapter 4 by taking the **ACE quizzes** on the student web site.

Chapter 5

1. Boyle, Charles, and Avogadro studied the relationship between two variables of a gas while keeping the other two variables that affect gases constant. What variables are studied and which variables are constant for Boyle's law, Charles's law, and Avogadro's law? Check your answers by playing the Boyle's Law and Charles's Law **Visualizations** on the student CD. Boyle's law can be depicted graphically in a few ways other than V vs 1/P. What does a plot of P vs V look like? PV vs P? How does each plot illustrate the inverse relationship between pressure and volume of a gas at constant moles and temperature?

2. Play the Liquid Nitrogen and Balloons **Visualization** on the student CD. Explain why the balloon collapsed when liquid N_2 was poured on the balloon in terms of the force of collisions per unit area. Why did the balloon return to its original shape at the end of the video?

3. Play the Ideal Gas Law, PV = nRT **Visualization** to review the relationships between P, V, n, and T. For each experiment, explain why the relationship is true in terms of the force of collisions per unit area.

4. Play the Collapsing Can **Visualization** on the student CD. Assuming only $H_2O(g)$ is in the can, what is the pressure of water vapor inside the can before the can is stoppered? When the can is immersed in water, what happens to the temperature of the gas molecules inside the can? What happens to the moles of gas molecules inside the can as the can is immersed? What affect does this have on P_{inside}? Put all these idea together and explain why the can collapsed when it was immersed in water.

5. What is diffusion? What is effusion? How is effusion of a gas related to the average velocity of the gas? To the molar mass of the gas? Check your answers by playing the Diffusion of Gases and the Effusion of a Gas **Visualizations** on the student CD. In terms of molar mass and average velocity, why did the $NH_3(g)$ and $HCl(g)$ molecules meet on the HCl side of the tube?

6. Test your understanding of Chapter 5 material by taking the **ACE quizzes** on the student web site.

Chapter 6

1. Read the Chapter 6 **Overview** on the student CD to review important topics covered in Chapter 6. Then, open the Thermite Reaction **Visualization**, read the concept, and play the video of this very exothermic reaction.

(a) Do the reactants or products in the thermite reaction have a higher potential energy? On average, do the reactants or products have the stronger bonds in this reaction? Explain how the potential energy difference between reactants and products is related to the heat produced in an exothermic reaction.

(b) Consider the following endothermic reaction: $N_2(g) + O_2(g) \rightarrow 2\ NO(g)$. Characterize this reaction in terms of potential energy differences and average bond strength differences between reactant and products. Where does the heat absorbed come from for an endothermic reaction? See Figure 6.3 in the text for a potential energy diagram for this reaction.

(c) The heat flow of a system at constant pressure is equal to the enthalpy change. For a chemical reaction, what are the signs of ΔH for an exothermic and for an endothermic reaction? Do Exercises 6.31 and 6.32 in the text to practice predicting signs of ΔH for some processes.

(d) The heat absorbed or released is a stoichiometric quantity related to the coefficients of the balanced equation. Do Exercise 6.33 or 6.34 in the text to practice determining quantities of heat released or absorbed in a chemical reaction.

2. The Thermite Reaction **Visualization** on the student CD shows a very exothermic reaction. Calculate ΔH for the thermite reaction using the Standard Enthalpy of Formation Values table. The reactants of the reaction are aluminum and iron(III) oxide; assume that the products are aluminum oxide and iron. For more practice on manipulating standard enthalpy of formation values, try Exercises 6.61, 6.65, and 6.67 in the text.

3. Test your understanding of Chapter 6 material by taking the **ACE quizzes** on the student web site.

Chapter 7

1. (a) Play the Electromagnetic Wave **Visualization** on the student CD. Are wavelength and frequency directly or inversely related? Next, view the Refraction of White Light **Visualization** to see the visible region of electromagnetic radiation. Does red light have a longer or shorter wavelength than violet light? Does yellow light have a lower or higher frequency than blue light? Which color of light travels at the faster speed? Review Figure 7.2 in the text to help answer these questions.

(b) The energy of electromagnetic radiation is transferred in units called photons as proposed by Albert Einstein. Open the Photoelectric Effect (Yellow or Blue) **Visualization** on the student CD to view the experiment whose results led Einstein to propose the idea of photons. Is the energy of a photon of light directly or inversely related to wavelength? Which color light, yellow or blue, represents the higher frequency light? Higher energy light? Longer wavelength light?

(c) The other types of electromagnetic radiation (in order of increasing wavelength) are gamma rays, x-rays, ultraviolet, visible, infrared, microwaves and radiowaves. Which is the most energetic type of electromagnetic radiation? Which type of electromagnetic radiation travels at the slowest speed: microwaves or visible light?

2. The great success of the Bohr model was its ability to predict the exact wavelengths of light emitted for the hydrogen emission spectrum. Play the Refraction of White Light and the H_2 Line Spectrum **Visualizations**.

(a) Using the Bohr model, why aren't all wavelengths of visible light emitted in the H_2 spectrum? Which line in the hydrogen emission spectrum is of highest energy?

(b) The four lines in the hydrogen emission spectrum correspond to the $n = 6$ to $n = 2$, $n = 5$ to $n = 2$, $n = 4$ to $n = 2$, and $n = 3$ to $n = 2$ electronic transitions in the Bohr model. Match the electronic transitions to the wavelengths listed in the H_2 Line spectrum animation. Verify that these four electronic transitions produce emissions at the four wavelengths listed in the **Visualization**.

3. View the Flame Tests **Visualization**. Each salt tested gives off a very distinctive color of light when burned. Explain why we see the flame turn a different color when the various salts are burned. Where is the light coming from? What is the purpose of the flame? Why do the various salts give off different colors of light when burned?

4. The shape of an orbital represents with 90% probability where the electron can be in that specific orbital. Why can't one be 100% sure where the electron is in a specific orbital? Describe the shapes and orientation of the s, p, and d orbitals. Check your answers by viewing the CD **Visualizations** on the various s, p, and d orbitals.

5. Test your understanding of Chapter 7 material by taking the **ACE quizzes** on the student web site.

Chapter 8

1. Work through the Formation of Ionic Compounds from Ions **Understanding Concepts** activity on the student CD. Complete the exercises until you are proficient at predicting formulas.

(a) In the exercises, the charges of the ions are given. However, you should be able to predict the cation charges for some of the metal ions given and the anion charges for all of the nonmetal ions given. Review Table 8.3 to

figure out how and why. Do Exercises 8.37 and 8.38 in the text to reinforce predicting ion charges.
(b) How is the size of a cation related to the size of the neutral atom? Explain. How is the size of an anion related to the size of the neutral atom? What is an isoelectronic series of ions? What determines the size trend in an isoelectronic series? Practice placing ions in order of increasing size by doing Exercises 8.33 and 8.35 in the text.

2. At the heart of the localized electron (LE) model is drawing Lewis structures. You must be proficient at drawing correct Lewis structures in order to apply the LE model. This takes practice.

(a) Once you feel comfortable drawing Lewis structures, open The VSEPR Theory of Molecular Structure **Understanding Concepts** activity on the student CD and skip directly to the Exercises. For now, just concentrate on drawing correct Lewis structures, so for each molecule or ion, draw the Lewis structure then check your answer by clicking Hint: Examine Lewis Structure.
(b) The Lewis structures in the Hint are incorrect for CO_2, NO_2^-, O_3, and SO_2. What was wrong with the answers? NO_2^-, O_3, and SO_2 all exhibit resonance. What is resonance and how many resonance structures can be drawn for these three species? How many resonance structures can be drawn for CO_3^{2-}? for SO_3?
(c) There were several exceptions to the octet rule in the various choices. Which two second-row atoms often have fewer than eight electrons? Which elements sometimes have more than 8 electrons?

3. (a) Open The VSEPR Theory of Molecular Structure **Understanding Concepts** activity on the student CD. The table at the end of the example offers an excellent review of all the possible shapes. Review the table to make sure that you can list approximate bond angles present for each shape. For example, see-saw shaped molecules would exhibit approximate 90° and 120° bond angles, whereas T-shaped molecules would exhibit approximate 90° bond angles (no 120° bond angle exists anymore). Bond angles are sometimes smaller than predicted from the VSEPR model. Explain this phenomenon.
(b) Once you are comfortable with the various shapes in the table, go to Exercises in the activity and predict the shape for every example available. This is an excellent review exercise on shapes. While going through each molecule/ion, also predict the approximate bond angles.

4. Test your understanding of Chapter 8 material by taking the **ACE quizzes** on the student web site.

Chapter 9

1. Open the Chapter 9 **Visualizations** on the student CD and view the sp^3, sp^2 and sp hybridization material. Which hybridization is always used for a tetrahedral arrangement of electron pairs? For a trigonal planar arrangement? For a linear arrangement? What's wrong with the statement "The methane molecule (CH_4) is a tetrahedral molecule because it is sp^3 hybridized"?

2. View the s-orbitals/bonding and antibonding **Visualization** on the student CD. Why is the bonding orbital lower in energy as compared to the separate atoms? Why is the antibonding orbital higher in energy? Open the Molecular Orbitals in H_2 **Understanding Concepts** activity and read through the introduction. What would be the bond order for the H_2^+ ion? For the H_2^- ion? Using MO theory, explain why He_2 molecules don't form.

3. (a) When MO theory is applied to Row 2 and heavier elements, the *p* orbitals are used to form molecular orbitals. One type of molecular orbital that can form from *p* orbitals is π orbitals. Open the Molecular Orbitals in H_2 **Understanding Concepts** activity and read about π bonding and π antibonding molecular orbitals. If the x-axis is the internuclear axis (the axis that goes through both nuclei of a diatomic molecule), which *p* orbitals form the π bonding and π antibonding orbitals?

4. What is a paramagnetic molecule? A diamagnetic molecule? What do you look for in the molecular orbital electron configuration to determine the magnetic properties? Is N_2 paramagnetic or diamagnetic? How about O_2? Check your answer by opening the Magnetic Properties of Liquid Nitrogen and Oxygen **Visualization**. The fact that B_2 is paramagnetic results in the changing of the expected molecular orbital energy level diagram. Explain why this was necessary.

5. Draw all the possible resonance structures for NO_3^-. What is the hybridization of the central nitrogen atom? Open the Pi Bond **Visualization** to view the π bonding system in NO_3^-. All the nitrogen-oxygen bonds are equivalent in NO_3^-. Would you expect the N-O bond strength to be that of a single bond on a double bond or something else? Explain.

6. Test your understanding of Chapter 9 material by taking the **ACE quizzes** on the student web site.

Chapter 10

1. (a) Open the Intermolecular Forces **Understanding Concepts** activity on the student CD to review types of intermolecular forces for covalent compounds. What type of intermolecular forces do nonpolar covalent compounds exhibit? How is size related to the strength of London dispersion forces? An extremely strong type of dipole-dipole force is hydrogen bonding. What three covalent bonds do you look for in order for a covalent compound to exhibit hydrogen bonding? How does the strength of ionic forces compare to hydrogen-bonding?

(b) Practice predicting the strongest type of intermolecular forces for various covalent compounds by doing the Exercises in the activity. Predicting whether the compounds are polar or nonpolar is an important part of this exercise that requires you to predict molecular structures.

2. Crystalline solids have a highly ordered structure. The smallest repeating unit in a crystalline solid is called the unit cell. Open the Unit Cells and Crystal Packing **Visualization** on the student CD to review three common types of cubic unit cells.

(a) When discussing metals, we commonly assume that the metal atoms are uniform, hard spheres that occupy the lattice points of the unit cells. For each cubic unit cell, how many net spheres (atoms) are contained in a unit cell?

(b) The atoms in a simple cubic unit cell are assumed to touch along the edges of the unit cell. With this in mind, how is the radius of an atom related to the edge length of the unit cell? Answer this same question for a body-centered unit cell and for a face-centered cubic unit cell. (Hint: The atoms are assumed to touch along the body diagonal in a body-centered cubic unit cell and to touch along the face diagonal in a face-centered cubic unit cell. You will also need to apply the Pythagorean theorem.)

(c) A common application of the relationships discussed above is to estimate the density of various metals knowing the type of unit cell. Do Exercises 10.49 and 10.52 in the text to examine some typical problems of this nature. A closest packed structure has the atoms packed together in the most efficient way. What is the difference between hexagonal closest packing and cubic closest packing? Which cubic unit cell is present in cubic closest packing?

3. (a) Not all solids pack in a very ordered array. Amorphous solids show considerable disorder in their structure. Glass is a common example of an amorphous solid (see Figure 10.28 in the text). The primary component of glass is silica, which has an empirical formula of SiO_2. From the formula of silica, one might expect SiO_2 and CO_2 to have similar structures; however, this is not the case. Compare and contrast the structure of CO_2 to that of silica.

(b) Ceramics can also be based on silicon-oxygen bridged compounds (called silicates). However, other chemical forms of ceramics exist including materials made from oxygen and metals. An important property of some oxide ceramics is superconductivity. View the Magnetic Levitation by a Superconductor **Visualization** on the student CD to learn more about superconductivity. Unlike glass, these superconductivity ceramics are crystalline in nature. To learn more about the crystalline structures of these superconducting ceramics, do Exercises 10.73 and 10.74 in the text.

4. Open the Changes of State **Understanding Concepts** activity on the student CD to review phase changes and heating curves.

(a) When a solid converts to a liquid and then to a gas, what happens in terms of the intermolecular forces? Rationalize why ΔH_{vap} for water is over 6 times larger than ΔH_{fus}.

(b) Go to the page with the heating curve. The heating curve has 5 basic parts to it. There are two plateau regions and three positive sloping lines in a typical heating curve. To calculate the energy necessary to convert some ice at a certain temperature below 0°C to gas at a temperature above 100°C, five quantities must be known: ΔH_{vap}, ΔH_{fus}, the specific heat capacity of gaseous water, the specific heat capacity of water, and the specific heat capacity of ice. Match these five quantities to the five parts in a typical heating curve. For H_2O, what temperature is the plateau at the lower temperature? The plateau at the higher energy? Do Exercise 10.83 in the text, which illustrates this type of calculation. Then do Exercise 10.81 in the text to test your ability to draw heating curves. Finally, try Exercises 10.85 and 10.100 for some variations on phase change energy calculations.

(c) Heating curves only represent substances at one particular pressure (typically 1 atm). Phase diagrams represent stable phases of a substance at many temperatures and pressures. Sketch the phase diagram for water. What happens to the melting point of water as the external pressure increases? What happens to the boiling point as the external pressure increases? For more typical phase diagram questions, do Exercises 10.87, 10.88, and 10.89 in the text.

5. Test your understanding of Chapter 10 material by taking the **ACE quizzes** on the student web site.

Chapter 11

1. (a) Open the Dissolution of a Liquid in a Solid **Visualization** on the student CD to review the hydration process for an ionic compound. To calculate the heat of a solution (ΔH_{soln}) when an ionic compound dissolves in water, you need to know the lattice energy of the ionic compound and the enthalpy change when the gaseous ions are hydrated (called the heat of hydration, ΔH_{hyd}).

(b) Using NaCl as an example, write out the chemical equation where ΔH equals the lattice energy. Write out the equation for ΔH_{hyd} and ΔH_{soln}.

(c) Using Hess's law, show how to manipulate the lattice energy equation and the ΔH_{hyd} equation to come up with the ΔH_{soln} equation. If ΔH_{soln} for NaCl is 3 kJ/mol and the lattice energy for NaCl is -786 kJ/mol, determine ΔH_{hyd} for NaCl. For more practice with this type of problem, do Exercises 11.33-11.36 in the text. Do Exercise 11.77 to review calorimetry problems.

2. Test your understanding of Chapter 11 material by taking the **ACE quizzes** on the student web site.

Chapter 12

1. (a) View the Reaction Rate and Concentration **Visualization** on the student CD. Is the order of HCl in the rate law a positive or a negative value (or equal to zero)? Explain how you know.
(b) The rate law for a reaction must be determined from experiment (not from the coefficients in a balanced equation). One way to determine the rate law is by using the method of initial rates. Explain how this method works.
(c) Consider the reaction between NO(g) and F_2(g). If doubling the concentration of NO (while keeping [F_2] constant) causes the initial rate to increase by a factor of 2, then what is the order of NO in the rate law? From experiment, the order of F_2 in the rate law is 1. If the rate increases by a factor of 4, what happened to the concentration of F_2 if [NO] was constant? If in a different reaction, tripling the concentration of some substance causes the rate to increase by a factor of nine, what would be the order of this substance? What is the relationship between concentration and rate for a zero order reaction? For a reactant with -1 as the order?

2. (a) Most chemical reactions occur by a series of steps called a reaction mechanism. Open the Oscillating Reaction **Visualization** on the student CD to see an example of this. For simplicity, let's assume this is a two-step reaction. Step 1 is the formation of the blue colored solution and Step 2 is the formation of the clear colored solution. Which of these steps would be the rate determining step in the mechanism? What is a rate determining step?
(b) To review the various ideas associated with a mechanism, let's consider the following two-step mechanism for the depletion of ozone by NO:
Step 1   NO(g) + O_3(g) → NO_2(g) + O_2(g)
Step 2   O(g) + NO_2(g) → NO(g) + O_2(g)
What species is the intermediate? What species is the catalyst? What is a catalyst? What is the overall balanced equation? If Step 1 is the rate determining step, what is the rate law derived from this mechanism? If you are having trouble answering these questions, review Sections 12.6 and 12.8 in the text. For some more problems with mechanisms, do Exercises 12.45, 12.47, and 12.48 in the text.

3. For an introduction to the collision model, open The Gas Phase Reaction of NO and Cl_2 **Visualization** on the student CD. The product of the reaction is ClNO. What color sphere is N in the animation? O? Cl? The animation shows that only collisions of specific orientation result in product formation. Which collision orientation results in product formation? Give an example of a collision that would not result in product formation.

4. For a more detailed review of the collision model, open the Transition States and Activation Energy **Understanding Concepts** activity on the student CD.

(a) Review the general reaction coordinate diagram for an endothermic reaction in the activity. Sketch a general reaction coordinate

17

diagram for an exothermic reaction indicating ΔH_{rxn}, $E_{a,f}$ and $E_{a,r}$. Is $E_{a,f}$ greater then or less than $E_{a,r}$ for an exothermic reaction? With all concentrations equal, which is slower, the forward or the reverse reaction for an exothermic reaction?

(b) Click More Examples in the activity to see reaction coordinate diagrams for several reactions. For each example, determine ΔH_{rxn} from the activation energies given. For all these reactions, the rate of the forward reaction will increase as temperatures increases. Why?

(c) Why does a catalyst speed up a reaction? Assuming the decomposition of C_2H_5Cl reaction can be catalyzed, sketch what the reaction coordinate diagram would look like for the catalyzed and uncatalyzed reaction. Show all the activation energies. Did ΔH_{rxn} change for the catalyzed reaction? For more practice with reaction coordinate diagrams, do Exercises 12.51 and 12.52 in the text.

4. Test your understanding of Chapter 12 material by taking the ACE quizzes on the student web site.

Chapter 13

1. (a) Open the Equilibrium Decomposition of N_2O_4 **Visualization** on the student CD. Equilibrium is a dynamic process. What does this mean? For a reaction at equilibrium, what is true about the rate of the forward and reverse reactions and what is true about the concentrations of reactants and products?

(b) The **Visualization** examines the N_2O_4 equilibrium as well as the effect of temperature on this equilibrium. For the reaction $N_2O_4(g) \Rightarrow 2\ NO_2(g)$, what is the equilibrium constant expression (K) in terms of molarity concentration units? What is the equilibrium constant expression (K_p) in terms of partial pressures? As long as temperature remains constant, the values of K and K_p remain constant no matter what initial concentrations of reactants or products are used. However, when temperature changes, the values of K and K_p change. From the N_2O_4 animation, did the value of K (or K_p) increase or decrease as temperature increased?
Hint: Compare the amount of NO_2 and N_2O_4 present at equilibrium before and after the heating.

2. (a) Open the Equilibrium **Understanding Concepts** activity on the student CD and read the introduction to review Le Chatelier's principle. When a reactant or product is added, the reaction shifts to reestablish equilibrium. Does the value of K change when a reactant or product is added? For a gaseous reaction, does the value of K change when the container volume is changed? Consider the reaction $N_2O_4(g) \Rightarrow 2\ NO_2(g)$. If the total pressure of the system at equilibrium is doubled by halving the volume, which way will the reaction shift to reestablish equilibrium? If the total pressure of the system at equilibrium is doubled by adding an inert gas like He, what effect does this have on the equilibrium?

(b) Next, go through the example, which examines the Fe^{3+} + NCS^- $\Rightarrow$ $FeNCS^{2+}$ equilibrium. The exercises study the effect of adding several reagents to this equilibrium. Your job is to use Le Chatelier's principle to predict which way the reaction shifts with each reagent added. It's important to know the color of solution when the hydrated Fe^{3+} dominates and the color of solution when $FeNCS^{2+}$ dominates. What color is each of these species?

(c) Go through all eight exercises and predict which way the reaction shifts to reestablish equilibrium. Also predict the color change that should occur. Play the reaction to verify your answer and choose Why to see a detailed explanation. The effects that some of the reagent additions will have are not obvious.

Here are some hints to help you deduce what should happen:
Addition of tin(II) chloride: Sn^{2+} reacts with Fe^{3+} to produce Sn^{4+} and Fe^{2+}.
Addition of silver nitrate: Ag^+ reacts with NCS^- to produce solid AgNCS.
Addition of sodium hydrogen phosphate: HPO_4^{2-} reacts with Fe^{3+} to produce $FeHPO_4^+$.
Addition of ammonia: The weak base NH_3 produces OH^- which reacts with Fe^{3+} to produce solid $Fe(OH)_3$.
Addition of Heat: The reaction is exothermic.
(d) The value of K changes for a reaction when temperature changes. For an exothermic reaction, does K increase or decrease when temperature increases? For an endothermic reaction, does K increase or decrease when temperature increases? For more practice with Le Chatelier's principle, do Exercises 13.59, 13.61, 13.63, and 13.64 in the text.

3. Test your understanding of Chapter 13 material by taking the **ACE quizzes** on the student web site.

Chapter 14

1. Open the Weak and Strong Acid **Understanding Concepts** activity on the student CD to review differences between strong acids and weak acids. The exercises in this activity allow you to compare major species present at equilibrium for two different acids and then answer some questions regarding relative acid strength between the two acids. Do several of the exercises until you are proficient at relating major species present to relative acid strength. For the pH question, the solution with the higher pH will be the weaker acid (the acid that has the fewest molecules disassociated).

2. Another way to differentiate strong acids from weak acids is by conductivity experiments. Open the Conductivities in Aqueous Solutions **Visualization**. Knowing that acetic acid is a weak acid and HCl is a strong acid, predict the relative brightness of the light for the acetic acid solution vs. the HCl solution. Play the video to confirm your answer. In terms of major species present at equilibrium, explain why the HCl solution produced a brighter light as compared to the acetic solution.

3. (a) Open the Self-Ionization of Water **Visualization**. The ion product constant, K_w, refers to a specific reaction. What is the K_w reaction? H^+ is commonly used as an abbreviation for what substance? What is the value of K_w at 25°C? Using the value of K_w at 25°C and the K_w expression, why does $[H^+] = [OH^-] = 1.0 \times 10^{-7}$ M in water when neither an acid nor a base is present? What is pH and pOH? Is pH and $[H^+]$ directly or inversely related? For a neutral solution of water, how is pH and pOH related at 25°C?
(b) When a lot of an acid is dissolved in water, how is $[H^+]$ related to $[OH^-]$ and how is pH related to pOH? How does the pH of a 0.10 M strong acid solution compare to the pH of a 0.10 M weak acid solution? When a lot of a base is added to water, how is $[OH^-]$ related to $[H^+]$ and how is pOH related to pH?

4. Test your understanding of Chapter 14 material by taking the **ACE quizzes** on the student web site.

Chapter 15

1. (a) Open the Buffers **Understanding Concepts** activity on the student CD. In the example, when 0.10 mol of HCl is added to the 1.0 M HF/1.0 M F^- buffer, the pH changes from 3.16 to 3.07. How does this illustrate a primary property of buffers? What would happen to the concentrations of 1.0 M HF and 1.0 M F^- if 0.10 mol of NaOH were added instead? Calculate the pH of the resulting

19

solution. In general, best buffers have a pH value close to the pK_a value of the weak acid component (pH = pK_a for best buffer). Using the Henderson-Hasselbalch equation, how are the weak acid concentration and the conjugate base concentration related for a best buffer?
(b) Under More Examples, HCl is added in 0.20 mol increments to pure water and to a 1.0 M HF/1.0 M NaF buffer. Notice the slow change in pH of the buffer as compared to the change in pH of water as HCl is added. This exemplifies the primary property of buffers as solutions that resist pH change. One thing you should be able to do is to calculate the pH as 0.20 mol increments are added. To practice, check that the pH values given are correct. Hint: For the buffer, added H^+ from the HCl can be assumed to react completely with F^- to form HF. Perform this stoichiometry problem first to determine the equilibrium concentrations of HF and F^- in the buffer, then do the buffer problem.

2. (a) Open the Neutralization of a Strong Acid by a Strong Base **Visualization** on the student CD, which illustrates what happens during the titration of a strong acid by a strong base. When HCl is added to water, what happens? Is the solution acidic, basic or neutral? As NaOH is added to the solution containing the HCl, what happens?
(b) Let's assume that 3 molecules of HCl are present initially. If 2 molecules of NaOH are added, what would be present in solution initially before any reaction? What would be present after the OH^- reacts completely with the H^+? Is the solution acidic, basic or neutral? If 3 molecules of NaOH are added, would the solution be acidic, basic or neutral? This is called the equivalence point and is represented in the animation. What is meant by the equivalence point? When 5 molecules of NaOH have been added, is the solution acidic, basic, or neutral? Explain in terms of major species present after the OH^- neutralizes (reacts) with the H^+ present initially.
(c) With the above discussion in mind, do Exercises 15.53 and 15.54 in the text, which illustrate strong acid/strong base titration problems. For each problem, sketch the pH curve and then compare and contrast the two curves. Make sure that you understand why the pH values are acidic, basic or neutral at the various volumes of titrant added. The key is realizing whether you have excess H^+ from the strong acid, excess OH^- from the strong base, or equal amounts of H^+ and OH^- (the equivalence point).

3. (a) Review Table 15.4 in the text, which lists K_{sp} values for various salts (ionic compounds). The K_{sp} values listed refer to a specific reaction. What is this reaction? Look at the fluoride salts. Which salt is most soluble in water? Least soluble in water? The K_{sp} value for AgCl is larger than the K_{sp} value for CaF_2, yet CaF_2 is more soluble in water than AgCl. Why is this? Calculate the molar solubility of each of these salts. Explain the terms saturated and supersaturated. View the Supersaturated Sodium Acetate **Visualization** on the student CD to learn more about supersaturated solutions.

4. Test your understanding of Chapter 15 material by taking the **ACE quizzes** on the student web site.

Chapter 16
1. (a) Read the **Overview** on the student CD for Chapter 16 to review important topics covered in this chapter. What is the second law of thermodynamics? What determines ΔS_{sys}? ΔS_{surr}? Practice predicting signs for ΔS_{sys} and ΔS_{surr} by doing Exercises 16.20, 16.21, and 16.29 in the text.

(b) How is ΔS_{surr} calculated? Why is the negative sign in the equation? Why is ΔS_{surr} inversely related to temperature?
(c) To predict spontaneity, ΔS_{sys} and ΔS_{surr} must both be considered. At what values of ΔS_{univ} is a process spontaneous? Under what sign combination of ΔS_{sys} and ΔS_{surr} is a process always spontaneous? Never spontaneous? Sometimes spontaneous? See Table 16.3 in the text to check your answers.

2. Test your understanding of Chapter 16 material by taking the **ACE quizzes** on the student web site.

Chapter 17
1. Open the Electrochemistry **Understanding Concepts** activity on the student CD and go through the introduction and examples to review redox reactions and galvanic (voltaic) cells.

(a) What is the purpose of the salt bridge? Cations in a salt bridge always flow to the cathode, whereas anions in a salt bridge always flow to the anode. In terms of the zinc-tin galvanic cell illustrated in the activity, explain why this is the case. Which way do electrons always flow in a galvanic cell?
(b) Standard reduction potentials are determined relative to some standard. What is the standard?
(c) Given two reduction potentials to design a galvanic cell, how do you determine which is the reduction half-reaction and which is the oxidation half-reaction?
(d) For additional review material on galvanic cells, open the Galvanic (Voltaic) Cells Visualization on the student CD.

2. To see an example of an electrolytic cell, open the Electrolysis of Water **Visualization** on the student CD.

(a) What half-reactions are occurring at the anode and the cathode? Hint: Review Table 17.1 in your text.
(b) What gas was produced on the right side of the apparatus? On the left side? The volume of gases produced are a ratio of about 2:1. Does this make sense?
Hint: Determine the overall reaction occurring when water is electrolyzed.
(c) Compare and contrast electrolytic cells and galvanic cells. What are some uses of electrolytic processes?

3. Test your understanding of Chapter 17 material by taking the **ACE quizzes** on the student web site.

Chapter 18
1. Chapter 18 introduces nuclear chemistry, including radioactive decay, balancing nuclear reactions, kinetics of nuclear decay, uses of nuclear decay, energy changes in nuclear reactions, nuclear fission and fusion reactions, and effects of radiation. Read the **Overview** on the student CD to review these topics.

2. Test your understanding of Chapter 18 material by taking the **ACE quizzes** on the student web site.

Chapter 19
1. Chapter 19 presents the Group 1A, 2A, 3A, and 4A elements. The material to emphasize for each group of elements includes the properties discussed, bonding characteristics, and typical reactions. Read the **Overview** on the student CD to review important topics covered in Chapter 19.

2. Open the Dry Ice and Magnesium **Visualization** on the student CD and view the reactions. Magnesium can react with oxygen as well as with carbon dioxide. Write balanced equations for these reactions.

Hint: As indicated in the concept, magnesium displaces oxygen in CO_2 forming magnesium oxide and carbon.
(a) What is oxidized and reduced in each reaction?
(b) Why did the match go out when placed in the beaker of CO_2? Sometimes reactions with magnesium can get out of hand. Why shouldn't you use a CO_2 fire extinguisher to put out a magnesium fire?

3. Test your understanding of Chapter 19 material by taking the **ACE quizzes** on the student web site.

Chapter 20

1. Chapter 20 presents the Group 5A, 6A, 7A, and 8A elements. As in Chapter 19, the material to emphasize for each group of elements includes the properties discussed, bonding characteristics, and typical reactions. Read the **Overview** on the student CD to review important topics covered in Chapter 20.

2. Chapter 20 emphasizes predicting shapes of the various covalent compounds formed by Group 5A-8A elements. To review predicting shapes of molecules, go back to chapter 8 on the student CD and work through The VSEPR Theory of Molecular Structure **Understanding Concepts** activity. The text uses the term distorted tetrahedron to describe the shape of XeO_2F_2. What term is used in the activity to describe the shape of XeO_2F_2?

3. The most important hydride of nitrogen is ammonia. What is the major use of NH_3? What are the acid-base properties of NH_3? Would you expect NH_3 to be soluble in water? Why? Go back to chapter 14 on the student CD and open the Ammonia Fountain **Visualization**. Explain why water fills the ammonia container in this demonstration.

Hint: Think about what happens to the pressure of NH_3 in the ammonia container as water is squirted into this container.

4. Test your understanding of Chapter 20 material by taking the **ACE quizzes** on the student web site.

Chapter 21

1. Chapter 21 discusses properties of transition metals with an emphasis on coordination compounds. For any coordination compound, make sure you can predict the structure, name the compound, draw possible isomers, and apply the crystal field model. Read the **Overview** on the student CD to review important topics in Chapter 21.

2. Open the Flame Tests **Visualization** on the student CD. The color of light in the various flame tests comes from the colors of visible light emitted as excited electrons drop down from higher to lower energy levels for the various ions. Notice that the compounds of copper used in the flame tests are colored (blue for copper (II) nitrate and green for copper (II) chloride), while the other nontransition metal compounds are white. Typically, transition metal complexes are colored.

One of the successes of the crystal field model is its ability to explain why transition metal complexes are colored. How does the crystal field model account for the colors of complex ions? Make sure you can do Exercises 21.44, 21.49, 21.53, which refer to the crystal field model.

3. Test your understanding of Chapter 21 material by taking the **ACE quizzes** on the student web site.

Chapter 22

1. Open the Synthesis of Nylon **Visualization** on the student CD. What type of polymer is nylon? Nylon is a copolymer. What does this mean? Review Figure 22.16 in the text for the structures of the monomers useful to produce the nylon in the video. What small molecule is eliminated when the nylon in the video forms? For more practice with condensation polymers, do Exercises 22.75 and 22.77 in the text.

2. Explain why carbohydrates are also polymers. Open the **Molecule Library** on the student website and view D-fructose and D-glucose to see two different monosaccharides.

3. Sucrose is a disaccharide. What does this mean and what two monosaccharides form sucrose? Open the **Molecule Library** on the student website and view sucrose. Starch, cellulose, and glycogen are polysaccharides. What does this mean? What monomer is used to produce starch, cellulose, and glycogen?

4. DNA and RNA are also polymers. However, the monomers used to form these polymers are more complicated than in proteins and carbohydrates. For DNA and RNA, each monomer contains three parts. What are the three parts? Open the **Molecule Library** on the student website and view the sugar part (deoxyribose in DNA and ribose in RNA) and the nitrogen-containing base part (uracil, cytosine, thymine, adenine, and guanine). What are the base pairs in the DNA double-helix structure?

5. Lipids are a class of substances that are water insoluble; they contain mostly carbon and hydrogen atoms. Does this make sense in view of lipids inability to dissolve in water? Fats are one type of lipid. Open the **Molecule Library** on the student website and view tristearin, the most common animal fat. Tristearin is a triglyceride.

6. Steroids are another type of lipid. Open the **Molecule Library** on the student website and view the steroids cholesterol, testosterone, progesterone, estradiol, and cholic acid. What common structural characteristic do all steroids have?

7. Test your understanding of Chapter 22 material by taking the **ACE quizzes** on the student web site.

HOUGHTON MIFFLIN COMPANY LICENSE AGREEMENT
Before opening this package, you should carefully read the following questions and answers regarding your use of the enclosed product. Opening the package indicates your acceptance of the terms and conditions contained in the answers. If you do not agree with them, you should return the package unopened and your money will be refunded.

Q: WHAT DOES THIS PRODUCT INCLUDE?
A: This product includes compact or floppy disks, software recorded on the disks, contents delivered by the software, and printed documentation.

Q: HOW MAY I USE THIS PRODUCT?
A: You may use the product as follows:
- You may only copy the software onto a single computer for use on that computer and you may only make one archival copy of the software for backup purposes only.
- You may print, transmit or modify the contents only for your individual use and use in the classroom. The product identifies any contents owned by rights holders other than Houghton Mifflin Company and you may only use those contents in accordance with current copyright law.
- You may not remove or obscure any copyright, trademark, proprietary rights, disclaimer, or warning notice included on or embedded in any part of the product.
- You may not reverse compile, reverse assemble, reverse engineer, or modify the software, or merge any portion of the software into another computer program.

Q: MAY I USE THE PRODUCT ON MORE THAN ONE COMPUTER?
A: You may use the product on more than one computer as long as there is no possibility that two different people will use the product on two different computers at the same time. If you want to use the product on more than one computer at a time, you must purchase separate copies for each computer location. If you represent an institution intending to use the product in an instructional computer lab, please refer to the networking provisions below.

Q: MAY I USE THE PRODUCT ON A NETWORK?
A: You may only use the product on a network if you represent an institution and have first purchased a site license from Houghton Mifflin Company. You may purchase a site license for the product by contacting our Customer Service department at 800-225-1464 with the proper ISBN and purchase order number from your department, lab, or school. Individuals wanting to use the product outside of an instructional computer lab must purchase separate copies.

Q: WHO OWNS THE PRODUCT?
A: You own the disk on which the software and the contents are recorded. Houghton Mifflin Company grants you a license to use the software and its contents in accordance with the terms and conditions set forth in this License Agreement. Houghton Mifflin Company and its licensors own and retain all title, copyright, trademark, and other proprietary rights in and to the software and its contents.

Q: MAY I GIVE THE PRODUCT TO ANOTHER PERSON?
A: You may transfer your license to use the product to another person as long as you permanently transfer the entire product (including all disks, all copies of the software program and all documentation provided in this package) without keeping a copy for yourself. To transfer your license properly, the recipient must first agree to the terms and conditions of this License Agreement. You may not otherwise license, sub-license, rent, or lease the product without permission from Houghton Mifflin Company.

Q: WHAT CAN I DO IF THE PRODUCT IS DEFECTIVE?
A: Within 30 days of purchase, contact Houghton Mifflin College Software at 800-732-3223 or e-mail: support@hmco.com. They will provide you with shipping instructions for the return and replacement of the defective disk. Provided the disk has not been physically damaged, Houghton Mifflin Company will replace the defective disk.
- LIMITED WARRANTY. EXCEPT AS EXPRESSLY STATED ABOVE, HOUGHTON MIFFLIN COMPANY MAKES NO WARRANTIES, EITHER EXPRESS OR IMPLIED, INCLUDING WITHOUT LIMITATION ANY WARRANTY OF MERCHANTABILITY OR FITNESS FOR A PARTICULAR PURPOSE.
- REMEDY. Your sole remedy is the replacement of a defective disk, as provided above.
- LIMITATION OF LIABILITY. IN NO EVENT SHALL HOUGHTON MIFFLIN COMPANY OR ANYONE ELSE WHO HAS BEEN INVOLVED IN THE CREATION, PRODUCTION, OR DELIVERY OF THE PRODUCT BE LIABLE FOR ANY INDIRECT, INCIDENTAL, OR CONSEQUENTIAL DAMAGES, SUCH AS, BUT NOT LIMITED TO, LOSS OF ANTICIPATED PROFITS, BENEFITS, USE, OR DATA RESULTING FROM THE USE OF THE PRODUCT, OR ARISING OUT OF ANY BREACH OF ANY WARRANTY.
- OTHER RIGHTS. Some states do not permit exclusion of implied warranties or exclusion of incidental or consequential damages. The above exclusions may not apply to you. This warranty provides you with specific legal rights. There may be other rights that you may have which vary from state to state.

Q: ARE THERE ANY RESTRICTIONS ON GOVERNMENT USE OF THE PRODUCT?
A: Houghton Mifflin Company provides this product to government agencies with restricted rights. Restrictions on government use, duplication and disclosure are set forth in subparagraph (c)(1)(ii) of the Rights in Technical Data and Computer Software clause at DFARS 252.227-7013 and subparagraph (c)(1) and (2) of the Commercial Computer Software - Restricted Rights clause at FAR 52-227-19.

If you have any questions about this License Agreement, call a Houghton Mifflin Contracts and Rights Analyst at (617) 351-3345.